SICKNESS INSURANCE

OR

SICKNESS PREVENTION?

RESEARCH REPORT NUMBER 6

MAY, 1918

NATIONAL INDUSTRIAL CONFERENCE BOARD

15 BEACON STREET

BOSTON, MASS.

Foreword

FEW problems are of more vital interest than the conservation of health. The public at large has only a faint conception of the extent to which sickness and physical disability sap the energies and impair the efficiency of the nation.

A broad, constructive program for dealing with this problem is a national need. Thus far, legislative attention has been directed largely toward compensation for the wage loss suffered by disabled workers rather than toward sickness prevention; yet the desirability of preventing sickness and its attendant losses is obvious.

Realizing the close relationship between national efficiency and health, the National Industrial Conference Board has attempted in this brief report to emphasize the enormous burden which sickness and physical disability place on society and industry and the imperative necessity for a thorough study of the practical possibilities of a program of sickness prevention with a view to its speedy adoption as a national policy. Yet this should be done without prejudice to the principle of compensation to the extent that its justification may be demonstrated. Obviously, however, with sickness prevention once established as a national policy, and with an efficient system of national health supervision, the necessity for resorting to some form of sickness insurance or compensation would be very greatly reduced.

Sickness Insurance
or
Sickness Prevention?

THE PROBLEM

Of 2,500,000 men examined for the National Army in 1917, 33% were found physically unfit and were rejected.[1] Of every 100 residents of the United States at least one is afflicted with tuberculosis in some form and a larger number with some organic affection of the heart. From 5 to 10% of adult males are estimated to have syphilis (either inherited or contracted) and the number indirectly affected by this insidious disease is much greater.[2] The proportions of those affected by minor ailments and defects are higher still.

That such conditions place a serious handicap upon the social well-being and productive efficiency of the nation is obvious. The practical question as to how they shall be dealt with demands most earnest consideration. Shall the effort of the state and of private management be directed chiefly toward prevention of sickness, or, instead, toward relief through treatment and compensation after disability arises?

In legislative discussion of the subject in this country chief emphasis has been laid on the desirability of compensation for sickness rather than on its prevention; yet the wisdom of preventing any loss which can economically be avoided is indisputable. Complete elimination of the losses arising from sickness is not practicable, but if a comprehensive program of sickness prevention can be made to yield substantial results from a reasonable expenditure, it should be adopted. A first step is to obtain a definite idea as to the extent of sickness and how far it is practically preventable. Then an estimate of the cost can be made and its justification determined.

[1] United States Provost Marshal General. Report to the Secretary of War on the First Draft under the Selective Service Act. 1917. pp. 44, 45.

[2] Some estimates run higher. See p. 7.

3

Comprehensive figures as to the extent of sickness are not available. A British statistician, Farr, has made an estimate that for every death there are, on the average, two persons constantly sick.[1] On the basis of the annual death rate in the United States in recent years — about 14 per 1,000 — such an estimate would indicate that about 3% of our population is constantly disabled by sickness and that on the average every man, woman, and child loses more than ten days per year through illness.

This estimate, however, seems rather high. The National Conference on Industrial Diseases, held in Chicago in 1910, estimated that a total of 284,750,000 days were lost through sickness by the 33,500,000 men, women, and children then engaged in gainful occupations in the United States, an average of eight and one-half days per worker.[2] This is in striking agreement with the results of sickness surveys made by the Metropolitan Life Insurance Company among policyholders and others in various localities. Seven such community surveys have been made by the company in widely separated localities which included a large variety of community types. A fair percentage of the total population of each place was studied, embracing, of course, a large proportion of the wage-earning population, and covering in most instances periods of either one or two weeks.

The percentages of those sick and the percentages of those who were so sick as to be unable to work are given in the following summary.

| | | PERCENTAGES | |
| | | | Unable |
COMMUNITY	SURVEY MADE IN	Sick	to work
Rochester, N. Y.	September, 1915	2.31	1.92
Trenton, N. J.	October, 1915	2.55	1.98
State of North Carolina	April, 1916	2.85	2.29
Boston, Mass.	July, 1916	1.96	1.80
Chelsea Neighborhood, N. Y. C.	April, 1917	1.48	1.38
Cities in Pennsylvania and West Virginia: (white and colored)	March, 1917	1.96	1.85
(colored)	March, 1917	2.31	2.18
Cities in Pennsylvania (white)	March, 1917	1.75	1.65
Cities in West Virginia (white)	March, 1917	3.30	3.11
Pittsburgh (white)	March, 1917	1.62	1.55
Kansas City, Mo.	April, 1917	2.52	2.39
Average, all surveys		2.02	1.88

[1] Irving Fisher. Report of the National Conservation Commission. Vol. III, p. 656.

[2] W. Gilman Thompson. The Occupational Diseases. p. 10.

4

This shows that on the average 2.02% of the population studied were sick and that 1.88% were unable to work, the proportions varying considerably in different localities.

With respect to the duration of illness, the Rochester survey showed that 50% of those who were sick had been incapacitated a year or more; the Boston survey showed 41.2%; and the surveys made in Pennsylvania cities, 24.7%. In Rochester 59.3% of individuals too sick to work were found to have been ill for more than 26 weeks; in Boston, 50.1%; in Pennsylvania (white persons), 28.9%; in North Carolina, 32.9%; in most of the other surveys, over 30%.[1]

The foregoing figures cover the entire population studied. For 376,573 persons 15 years of age and over covered in these surveys the average loss of time was 8.4 days per year, or 6.9 working days, based on 300 working days per year.[2] Females showed, on the average, a slightly higher rate of disability than males.

If the sickness rates disclosed by these surveys for persons 15 years of age and over hold for the industrial population of the entire country, the annual wage loss for 40,000,000 workers from sickness may be conservatively estimated at from $500,000,000 to $750,000,000.

PREVALENCE OF PHYSICAL DEFECTS AND DISABILITIES

Distinct from sickness proper but extremely important in their bearing on social welfare and national efficiency are the numerous defects and disabilities which impair productive capacity and which, in many cases, eventually result in disabling diseases. Mention already has been made of the high percentage of rejections among recruits to the National Army in 1917. Statistics based on portions of these rejections indicate the prominence of such physical disabilities as defective hearing and vision, decayed teeth, defects of the nose and throat, heart affections, flatfoot, and minor deformities — all of which constitute physical handicaps that must interfere

---

[1]The figures here given are taken from the reports on these community sickness surveys as published by the Metropolitan Life Insurance Company.

[2]The California Social Insurance Commission, in a report dated January 25, 1917, estimated that, among wage-earners in that state "an average of six days per person is lost each year because of sickness."

with maximum productive efficiency and some of which, if allowed to remain uncorrected, must sooner or later lead to serious illness or disability. Undoubtedly many of these defects could have been prevented if treated in early childhood.

A similar examination of the records of 72,410 applicants for service in the United States Navy for 1914 shows that 76% were rejected. Of the total number of rejections, 14.97% were due to defective vision, 8.61% to defective teeth, 10.77% to flatfoot, 7.48% to deformities, and 5.79% to heart affections. Study of the rejections from a group of applicants numbering 73,028 for the same service in 1915 showed practically the same causes and proportions as in 1914.[1]

Such figures disclose an unmistakably widespread prevalence of physical disability among a class of our population which should be in its prime. Surely, if the physical condition of these applicants disqualified them for service in the Navy, such conditions would constitute a handicap to them in other lines of work also.

FEEBLE-MINDEDNESS AND VENEREAL DISEASE

Overshadowing general physical defects and disabilities, but largely contributing to them, are two factors far-reaching in their results — feeble-mindedness and venereal disease.

It has been stated that there are "at least four feeble-minded persons to each thousand of the general population," and that feeble-mindedness produces more pauperism, degeneracy and crime than any other single force. From 25 to 50% of the inmates of our prisons and jails, from 15 to 30% of the inmates of almshouses, and a still larger percentage of prostitutes are attributed by some authorities to this class.[2] This statement, however, utterly fails to give an adequate conception of the widespread effects of feeble-mindedness. Feeble-minded-

---

[1]United States Bureau of Education. Annual Report, 1916. Vol. I, p. 318.

[2]H. H. Goddard. Feeble-mindedness, Its Causes and Consequences. pp. 7-9, 15, 17; Memorandum submitted to the Hospital Development Commission by the State Charities Aid Association and the New York Committee on Feeble-mindedness. New York. September, 1917. pp. 8, 13, 14; Helen MacMurchy. The Relation of Feeble-mindedness to Other Social Problems. In National Conference of Charities and Corrections. Proceedings, 1916. p. 233.

ness is particularly an inherited defect. The feeble-minded are peculiarly prolific, and the feeble-minded parent of today may therefore ultimately become responsible for a great number of public charges. This class is accountable also for an enormous amount of venereal infection and other sickness which, in an ever-expanding progression, exert their sinister effect on the national health.

Mental defectives and the mentally subnormal also contribute heavily to industrial inefficiency. Indeed, it is this group in particular that constitutes the inefficient, irresponsible, shifting workers of industry. While low paid, their inefficiency makes them the high-cost labor of industry.

Likewise of enormous importance in the burden which they throw on society and industry are the venereal diseases. For instance, to gonorrhea, more prevalent than any other disease except measles, are ascribed 6,000 to 10,000 cases of blindness annually (equal to 10% of all cases) and 80% of blindness in the new-born. Gonorrhea also results in many chronic diseases of both men and women and has been held responsible for the necessity of a very large percentage of all abdominal operations in the latter sex.[1]

Syphilis, as has been stated, is estimated to affect from 5 to 10% of the adult male population. The lower estimate is undoubtedly conservative.[2] Other estimates run as high as 15% for adult males. One puts the percentage at 8 for the total population. To syphilis also must be charged practically all locomotor ataxia[3] and 10% of all insanity.[4] Syphilis is, moreover, held responsible for a large proportion of the diseases of the heart, blood vessels, and other vital organs, and for more deaths than are caused by diphtheria, typhoid fever, scarlet fever, measles, whooping cough, and influenza combined.[5] It is charged with most of the deaths from apoplexy occurring before middle age.[6] Of the deaths from softening of the brain,

[1]American Social Hygiene Association. Venereal Diseases. Pamphlet. 1917. p. 15.

[2]William Allen Pusey. Syphilis as a Modern Problem. p. 106.

[3]American Social Hygiene Association. Venereal Diseases. p. 16.

[4]See Frankwood E. Williams. Relation of Alcohol and Syphilis to Mental Hygiene. In American Journal of Public Health. December, 1916. p. 1275.

[5]Massachusetts State Department of Health. First Annual Report of the Commissioner. 1915. p. 20.

[6]American Social Hygiene Association. Venereal Diseases. p. 16.

from general paralysis, from spinal cord diseases, and from cardiac and vascular diseases commonly occurring in the form of angina pectoris and arteriosclerosis, one in four is ultimately chargeable to this disease.[1]

Clearly, if the ravages of these insidious diseases and the burden of feeble-mindedness can be prevented it is of the utmost importance to prevent them.

VITAL CONCERN OF INDUSTRY IN THE PROBLEM

While the problem of sickness and physical disability affects the entire population, it is of particular concern to industry. In practically every branch of industry disease and disability cast their baneful influence. In the case of garment workers, for instance, studies by the U. S. Public Health Service show that nearly 3% of males between the age of 20 and 44 years were affected by arteriosclerosis, a similar proportion by kidney diseases and nearly 2% by valvular disease of the heart.[2] Similar studies of male food handlers for approximately the same age groups showed that about 7% had organic heart disease, 3½% diseases of the arteries, over 2% cirrhosis of the liver, and a similar percentage some form of kidney disease.[3]

When physical defects as well as disease are included, the proportions run very much higher. For instance, of 800 bakers examined in New York for the army and navy, 57% had some disease or defect; of a similar number of tailors the percentage was nearly 63;[4] of 203 printers and 1,600 food handlers, it was only a little below 70.[5] Of a group of 2,086 male garment workers practically 100% were affected by some disease or physical disability.[6]

[1]Albert Scott Warthin. A Plan for Combating Venereal Diseases in the State of Michigan. Michigan State Board of Health. Public Health. October-November, 1917. pp. 482, 483.

[2]U. S. Public Health Bulletin No. 71. p. 61.

[3]Department of Health of the City of New York. Monograph Series No. 17. August, 1917. Table facing p. 8.

[4]George M. Price. Occupational Diseases and the Physical Examination of Workers. In Fifteenth International Congress on Hygiene and Demography. Washington, 1912. Transactions, Vol. III, Part 2, p. 847.

[5]James A. Miller. Pulmonary Tuberculosis among Printers. 6th International Congress on Tuberculosis. Washington, 1908. Vol. III, pp. 210–213; Department of Health of the City of New York. Monograph Series No. 17. Table facing p. 8.

[6]U. S. Public Health Bulletin No. 71. pp. 54–61.

For such ailments as defective teeth, defective vision, diseased tonsils, rhinitis, spinal curvature, and flatfoot, numerous occupations show proportions ranging above 25%, sometimes above 50%.

To say that the existence of any such great amount of ill health and physical disability among the nation's industrial workers is a serious matter is merely to state a truism. Even though these disabilities may not, at least in their earlier stages, cause extended absence from work, the tax thus imposed on efficiency must be a heavy one. Obviously, maximum efficiency cannot be obtained from a force of workers one fourth of whom are suffering from such disabilities as defective vision, nasal disorders, and deformities, or whose health is being steadily sapped by tuberculosis, alcoholism, or venereal disease.

For most of the sickness and disability existing among the industrial population, industry cannot be held responsible. Undoubtedly some disease and disability have their origin within the factory. In the case of so-called occupational diseases the responsibility of industry is quite clear. Yet such occupational diseases constitute only a modest fraction of the total number of ills to which industrial workers are victims. For instance, of 5,121 cases studied in a special industrial clinic in the out-patient department of the Massachusetts General Hospital in 1916, in only 466 was a definite relationship established between the patient's defect or disability and the hazard incident to his work, although such relationship was indicated in a considerably larger number of instances. The 466 cases were comprised of the following: lead poisoning, 148; caisson disease, 12; "occupational strain," 91; industrial dermatoses, 54; anthrax, 18; occupational neuroses, 22; affections of the respiratory tract, 56; unstated, 65.[1]

Tuberculosis is often spoken of as a great industrial disease. Yet it is a very much disputed question as to how far tuberculosis is fairly chargeable to industry. In a vast number of cases responsibility for this disease must be borne in large measure by such influences as heredity, housing conditions, and personal habits. A direct connection between the prevalence of tuberculosis and congested housing conditions has been definitely established by many studies. Thus one survey made in Berlin showed

[1]Wade Wright. An Industrial Clinic. Monthly Review. U. S. Bureau of Labor Statistics. December, 1917. pp. 185–188.

that 42% of all cases of tuberculosis occurred in families occupying but one room, 40% among those occupying two rooms, 12% among those occupying three rooms, and only 6% among those occupying four or more rooms. Other influences, of course, undoubtedly were involved, but these statistics and many others of a similar character emphasize the importance of housing conditions as a factor in the tuberculosis problem.[1] The social and economic conditions underlying such housing congestion must, however, also be considered.

In the case of feeble-mindedness and venereal disease, industry clearly is the victim rather than the culprit. In an apportionment of the cost of dealing with these problems, whether by prevention or relief, this fundamental fact must be definitely recognized.

Furthermore, it should be appreciated that industry is concerned not alone with the sickness and physical disability of workers actually employed. For instance, an industrial manager who fancies that an excessive infant mortality or morbidity rate is no concern of his takes an extremely narrow view. It is from the infants of the present that the workers of the future must be recruited. Infant conservation is but another phase of industrial preparedness. Yet of all deaths in the United States, those of infants under one year of age comprise one-sixth,[2] and fully one-half of these are held to be preventable. In addition, a vast number of defects and ills which impair efficiency later in life have their origin in early childhood and could be prevented by proper treatment at that time. Such facts demonstrate that the interest of industrial managers in ill health and disability is by no means confined to that existing within the walls of their factories. They also demonstrate that these managers should take an active and intelligent part in the working out of an effective program of national health conservation.

How Far are Diseases and Physical Defects Preventable?

That a vast amount of disease is preventable, not only in theory, but as a practical measure and at a reasonable cost, cannot be questioned. Experience with such

---

[1] U. S. Public Health Bulletin, No. 76. Health Insurance: its Relation to Public Health. p. 23.
[2] United States Census. Mortality Statistics, 1915. p. 212.

diseases as typhoid fever, tuberculosis, pneumonia, and diphtheria affords convincing evidence that the opportunity for prevention, in the field of communicable diseases at least, is very great.

Within a forty-year period the death rate from typhoid fever has in many communities dropped from 50, 75, or more per 100,000 population to 10, and in some communities to less than 5. For the entire registration area of the United States Census the rate fell from 35.9 per 100,000 population in 1900 to 12.4 in 1915. The death rate from tuberculosis, which in 1900 was over 200 per 100,000 population for the registration area, has in recent years been cut to below 150.[1] Fully as striking are the results attained with diphtheria, and certain other contagious diseases. In view of such achievement, extended discussion to demonstrate the practicability of a preventive program in the field of communicable disease is superfluous.

From 1895 to 1916 the proportion of deaths from communicable diseases to total mortality in Massachusetts decreased from 28.6% to 18.48%, or at the rate of nearly one-half of 1% per year.[2] There seems to be no reasonable ground to doubt that broad success could also be attained in respect to many non-communicable diseases, provided the funds and a suitable organization were available.

In the course of a health survey made in one of the industrial centers of Massachusetts, physical examination was made of members of 720 families. The total number of examinations was 1,682. Major affections were found in 1,329 cases, tuberculosis in 48, while complicating illnesses or defects were reported in 1,211 cases, making the total number of affections discovered in the 1,682 examinations no less than 2,588. Of this total number of affections 55% were held to be preventable. The preventable and the doubtful groups together constituted 84% of the total number of defects, leaving only 16% in the definitely non-preventable group.[3]

The diseases most frequent in the industrial population, other than strictly "occupational diseases," are those

[1]United States Census. Mortality Statistics. 1915. pp. 21, 54, 58-142; see also Massachusetts State Department of Health. Annual Report. 1917. p. 221; New Jersey Department of Health. Public Health News. August, 1916. p. 6.

[2]Massachusetts State Department of Health. Annual Report, 1916. p. 199.

[3]D. B. Armstrong. Community Health and Tuberculosis Demonstration. Manuscript.

of the degenerative type, such as Bright's disease, cancer, apoplexy, cirrhosis of the liver, and arterial diseases. These are particularly frequent causes of disability and death after the age of 45 years, although they often exert a detrimental effect on the efficiency of workers at earlier ages. Predisposing factors which contribute to the occurrence of those diseases often exist in the very early periods of life. While opportunity for prevention of these diseases may be less promising than in the case of communicable diseases, there is undoubtedly a large field for successful preventive work if undertaken in time.

Many of these diseases are attributable in part to over-indulgence in alcoholic liquors and to the ravages of venereal infection. So far as alcoholism and venereal diseases are concerned, a preventive program must reckon with the factor of will power, which makes the problem of prevention more difficult. Undoubtedly such diseases will for a long time continue to tax society and industry. Nevertheless, even here the opportunity for valuable preventive effort cannot be questioned.

Enough has already been accomplished in the prevention of many diseases which have their origin in industrial poisons, skin irritants, fumes, vapors, and dust, to demonstrate the wisdom of more intensive prosecution of such work.

With respect to physical disabilities as distinct from sickness, the case for prevention is strong. There can be no question that many ailments, such as defective hearing or vision, nasal defects and many throat disorders, which impair efficiency even though they do not completely incapacitate the individual for work, can be entirely cured if treated in time. In many instances where complete cure cannot be accomplished a radical improvement certainly is possible.

To many diseases among workers of the larger industrial centers overcrowding, poor ventilation, and insufficient lighting are contributing factors. Such conditions clearly are preventable.

To fail to apply preventive measures to such illnesses, disabilities or conditions as will almost certainly respond, and instead to permit them to go uncorrected until the victim becomes a charge on society, is absurd. Certainly, if the state can contribute to the support of individuals after they become incapable of caring for themselves, it can contribute to prevent them from becoming incapacitated.

COMPARATIVE COST OF SICKNESS PREVENTION AND COMPENSATION

So far as preventable diseases and disabilities are concerned, the advisability of their prevention can be questioned only on the ground of cost. The problem in the case of such diseases and disabilities, therefore, is to determine whether their prevention would involve an expenditure out of proportion to the probable benefit.

Conservative estimates of the cost of sickness insurance (or health insurance,[1] as it is often called) set the total for the country at large at not less than $720,000,000 and perhaps not less than $1,000,000,000 per year. An expenditure of $720,000,000 per year means a per capita outlay of approximately $7 for each man, woman, and child in the country. In view of the undeniable fact that in nearly all programs of this sort the cost tends to increase as time goes on, there can be little question that the adoption of the compensation principle as a national policy would eventually mean a still larger addition to the taxpayer's burden. European experience with sickness insurance laws and our own experience with workmen's compensation laws clearly indicate that their cost after but a few years of operation usually greatly exceeds the preliminary estimates. The fact that the burden of an insurance program would be heavy, however, is not the fundamental reason for questioning its justification. The real consideration is the strong probability that the expenditure of only a modest fraction of this vast sum to combat disease, under an organized program of prevention, would avert an enormous amount of sickness and disability. The important fact already noted must not be lost sight of, viz.: that the greater part of the funds expended in preventive work, if wisely used, would be refunded to the community in the form of production which, under the insurance system, would be lost.

Obviously, the cost of timely correction of such minor disabilities as defective vision, defective teeth, minor

---

[1] In this report the term "sickness insurance" has been adopted.

deformities of posture, and a hundred and one defects of this class would be small in comparison with the loss which such disabilities might directly or indirectly entail if allowed to go unremedied.

It is important to bear in mind that no sickness or health insurance program yet offered purports to be available for the benefit of the entire community, or, in fact, for even all of the industrial population. Usually its benefits are restricted to manual workers, and to other wage-earners whose annual earnings fall below a certain limit. This difficulty, of course, can be met by extending the scope of sickness insurance. But if the limited application already proposed involves an outlay of three-quarters of a billion dollars or more annually, it is apparent that neither its adoption nor its extension is a matter to be lightly accepted.

The estimated cost of education in the United States in 1914, including private as well as public institutions, aggregated $754,500,000.[1] In other words, the educational facilities of the country are maintained at an estimated annual cost less than the estimated cost of sickness insurance. Education, however, is in no sense the counterpart of compensation, but is, rather, in the same class as preventive work. Obviously, society would not tolerate a proposal to foster a system whereby the education of its members would be neglected during youth, and instead, compensate by a system of cash benefits those who, because of a lack of education, would later be unable to compete successfully in the struggle for a livelihood. If it is good policy to safeguard the young by fitting them, through education, for their life work, is it not equally obvious that it is also sound policy to safeguard their health by early expenditure to prevent sickness and disability rather than to relieve the effects of such misfortune by belated expenditures for compensation?

EXPERIENCE OF EUROPEAN COUNTRIES WITH SICKNESS INSURANCE

The experience of those European countries which have adopted sickness insurance on a comprehensive scale is illuminating. A striking feature of this European experi-

---

[1] Department of the Interior. Report of the Commissioner of Education. 1916. Vol. II, p. 9. Of the total cost here given, $486,166,000 was for public elementary schools.

ence is that the average number of days lost on account of sickness, at least ostensibly, has steadily increased. In Germany, for instance, out of every 100 insured wage-earners in 1890, 36.7 were listed as sick at one time or another during the year; in 1913 the proportion was 45.6. In Austria, where sickness insurance is general, the corresponding figures were 45.7 in 1890 and 51.8 in 1913. Not only has the number of cases sharply increased, but the average number of days lost on account of sickness per sick member has likewise increased: in Germany, from 16.2 days in 1890 to 20.2 in 1913; in Austria during the same period, from 16.4 to 17.4.

Still more striking is the increase in the average number of days lost per member insured, which in Germany rose from 5.9 in 1885, when the sickness insurance laws had just gone into effect, to 6.19 in 1890 and to 9.19 in 1915, while corresponding Austrian statistics from 1890 to 1913 show an increase from 7.98 days per member insured to 9.45 days.[1]

It may fairly be argued that a temporary increase in number of days lost does not necessarily reflect a real increase in sickness. It might even indicate greater prevention by which more serious disability would be avoided. For instance, absence of a few days on account of an incipient cold might easily forestall an absence of several weeks because of pneumonia. But certainly, if this argument holds, the temporary increase in time lost should be, because of quick and thorough treatment, comparatively small, and should be checked, or followed eventually by a permanent decrease. The fact that in these countries the number of days lost through sickness has not decreased, but, on the contrary, has steadily and sharply increased over an extended period, discredits the suggestion that it is due to greater precaution. The most reasonable interpretation of this rise in the sickness curve in Germany and Austria under sickness insurance systems is that it reflects a growing tendency to malinger and take advantage of the sickness benefits provided. This is a serious and exceedingly practical criticism of the operation of the system which should receive most careful

---

[1] These figures quoted from Magnus W. Alexander. Some Vital Facts and Considerations in Respect to Compulsory Health Insurance. 1917. p. 11. See also Statistiches Jahrbuch für das Deutsche Reich. 1892. pp. 197, 198; United States Bureau of Labor Statistics. Monthly Review. May, 1916. p. 77; November, 1916. p. 127; W. Harbutt Dawson. Social Insurance in Germany, 1883–1911. pp. 91, 94.

attention in considering the wisdom of introducing it in the United States. Indeed, the history of European sickness insurance disbursements strongly suggests a widespread and growing disposition on the part of the unscrupulous to profit at the expense of the honest. Aside from the question of cost and fairness, this is vitally important because of its effect on the morale of industrial workers.

In significant contrast to this experience of Germany with compulsory health insurance is the record of mutual societies in France which showed a rather noteworthy decrease in the average days lost, while in Germany the average was steadily rising. Thus, for the so-called free societies of France (i.e., those having no subsidy from the Government) the average number of days lost through sickness per insured member fell from 6.23 in 1898 to 3.87 in 1910,[1] whereas in Germany the average rose from 6.16 to 8.53 (in 1911). In view of the fact that, under the mutual system, there is much less incentive for workers to malinger, this radical difference in the experience of the French mutual societies, as compared with the compulsory system of Germany, seems to give added force to the suggestion of malingering in the latter country. That malingering is common under the German system is, indeed, charged by many German authorities.

Again, the experience of Great Britain with its National Health Insurance Act has by no means been wholly satisfactory. It is true that the British Medical Society in reviewing the first five years operation of the Act took the ground that on the whole the results had been favorable and that the defects in the Act have been largely concerned with minor administrative details. On the other hand, the Act has been sharply criticized by others, particularly with respect to disappointing results obtained in sanatorium work in the case of tuberculosis. The following criticism of one authority seems pertinent:[2]

> "The fundamental fault in the national insurance act, from the point of view of public health, is that it does little or nothing to touch the great environmental causes of disease. It is palliative rather than preventive."

[1] H. G. Villard. Workmen's Compensation and Insurance in France, Holland, and Switzerland. 1914. p. 37.

[2] William A. Brend, in The 19th Century and After. July, 1917.

16

This statement, which is merely an opinion, is not presented here as proof that the National Health Insurance Act in Great Britain has been a failure. In view, however, of the obvious difference in opinion as to the results which have been achieved under that Act it is evident that a careful investigation into its operation should be made before accepting the British legislation as a basis for action in this country. It may be noted that the cost of health insurance in Great Britain has very greatly exceeded the preliminary estimates.

At the present time the establishment of a Ministry of Health is being vigorously advocated in Great Britain. Such a department, if organized, would devote much effort to the prevention of sickness.

EXPERIENCE OF NEW ZEALAND AND AUSTRALIA

The comparative experience of New Zealand and Australia with the infant mortality problem is likewise pertinent.

In New Zealand a vigorous educational campaign has been carried on by the Government, which through the agency of women's and children's societies, and the establishment of women's and children's hospitals, has made it possible for a large proportion of mothers in both urban and rural communities to obtain advice, nursing, and medical and hospital care. In 1911 the infant death rate in New Zealand was 56.31 per 1,000 births. In 1915 it had fallen to 50.05 per 1,000 births, the lowest infant mortality rate in the world.

In Australia, however, the Government grants a maternity allowance of £5 for each child. These government grants have been almost universally accepted and the total expenditure in 1916 reached £662,035. Nevertheless, in 36% of all births the mothers were not attended by a physician. The infant mortality rate in Australia was but slightly reduced, falling only from 68.49 per 1,000 births in 1911 to 67.52 in 1915.

While the great excess in the Australian rate over that in New Zealand may not be entirely attributable to the difference in methods of handling the infant mortality problem, the general similarity of conditions in the two countries gives a striking significance to the question of method. This is furthermore pertinently suggested by the fact that a special committee, appointed by the

17

Australian Commonwealth to investigate the problem of infant mortality, reported in June, 1917, to the Australian Parliament that there was urgent necessity for the adoption of measures in Australia similar to those so successfully applied in New Zealand.

A second committee, reporting on the subject in August of the same year, called attention to the fact that, although there had been a slight reduction in the infant death rate in Australia since the introduction of the cash benefit system, the decrease was, nevertheless, smaller than in the preceding year, in which no cash benefits were paid.[1]

PREVENTION THE ANTITHESIS OF INSURANCE

It has often been claimed that a sickness insurance system creates a new economic incentive for preventive work. The experience of the European countries just referred to does not support this contention. Indeed it is difficult to see any logical ground for the claim; a clear appreciation of the extent of sickness and disability and the heavy burden which they place upon society should be the sufficient and powerful incentive for prevention. Certainly that incentive gains no force by confusing it with insurance, the very antithesis of prevention. If interest in prevention can be aroused through an insurance system, it should be much more sharply stimulated by an organized program having prevention for its chief object. Furthermore, it is obvious that the incentive for a community to spend large sums in preventive work is not increased by first draining its resources to support an expensive system of treatment and insurance.

As contrasted with prevention, the chief aim of sickness insurance is to relieve the worker from the effects of, or to compensate him for the loss resulting from, sickness or disability already incurred. In most legislation of this sort the proposal is that the victim shall be compensated to the extent of a portion of his wages for a definite period. In the so-called model bill prepared by the American Association for Labor Legislation, which has been used in drafting sickness insurance measures in various states, the plan is to pay the sufferer two-thirds of his wages for a period not in excess of twenty-six weeks in any one year. This, of course, is a very definite measure of relief and one which the wage-earner can readily under-

---

[1]United States Children's Bureau. Fifth Annual Report. 1917. pp. 45–57.

18

stand; the payments proposed are the real basis for most of the agitation in favor of sickness insurance. If, however, the wage-earner, instead of being compensated over a period of two, ten, or twenty-six weeks to the extent of two-thirds of his wages, can be saved the disability and consequent loss of time by one-half this outlay, or even by an equal expenditure, it is clear not only that he is himself directly benefited but also that society as a whole secures an advantage; because, by prevention, the loss of production which would result if his disability were permitted to run into serious incapacitation is averted.

Present Attitude of State Legislators

It seems significant that there is a disposition on the part of legislators in the United States investigating the problem to go cautiously. Official investigating commissions have been appointed in nine states[1] to study sickness insurance. Four of these commissions (in three states) have made reports which are available for study.

The first Massachusetts Commission (1917) investigated not only sickness insurance, but also old-age pensions, unemployment, and hours of labor in continuous industries. Four of its nine members favored the adoption "of a general system of health insurance for wage-earners supported by enforced contributions from employers, employees, and the State." One member, while concurring in this recommendation, dissented as to the distribution of the cost until more "accurate information is available based on Massachusetts statistics." Two members urging their agreement with "the aim and purpose of health insurance to conserve the health of the wage-earner and his family" nevertheless opposed specific legislation "until the means by which this end may be attained are thoroughly understood and the public opinion is formed on the subject." The remaining members contended that "this Commission has not had sufficient time to study the subject thoroughly" and counseled against "immediate legislation."[2]

---

[1]California, Connecticut, Illinois, Massachusetts, New Hampshire, New Jersey, Ohio, Pennsylvania and Wisconsin.

[2]Massachusetts Special Commission on Social Insurance Report. February, 1917. pp. 22, 37, 42, 43.

The special Massachusetts Commission which reported in January, 1918, continued the study begun by the previous Commission. Of its eleven members, nine reported adversely. The two dissenting members favored the appointment of a new Commission for further investigation, and commended a non-contributory plan of insurance.[1]

The New Jersey Commissioners recommended the passage of a workmen's health insurance bill adapted to New Jersey's needs. They contended that special emphasis should be placed on prevention of sickness and expressed the opinion "that health insurance is a measure which gives great promise both of relieving economic distress due to sickness and of stimulating preventive action."[2]

The California Commission reported that "Health insurance of wage-earners would mean a tremendous step forward in social progress," but stated that 'it was not "prepared to offer a plan for the organization of health insurance." The Commissioners saw what they believed to be serious objections to the plan (that of the American Association of Labor Legislation) which had been offered. They sketched a plan of organization which they thought to be free from certain objections, but they conceded that this substitute plan might "be open to objections still more grave."[3]

ATTITUDE OF ORGANIZED LABOR

A number of labor organizations have expressed themselves in favor of the principle of health insurance, notably some of the international textile unions and the international typographical union. Several state federations of labor have likewise endorsed the principle of universal health insurance; in many cases, however, a non-contributory plan, which would throw the total cost either on employers alone or jointly on them and the state, was advocated. On the other hand, it may be noted that Mr. Samuel Gompers, President of the American Federa-

[1]Massachusetts Social Insurance Commission Report. January, 1918. pp. 36–55, 61–70.
[2]New Jersey Commission on Old Age Insurance and Pensions. Report on Health Insurance. pp. 2, 19.
[3]Report of the California Social Insurance Commission. January 25, 1917. pp. 23, 123.

tion of Labor, recently expressed himself as strongly opposed to the system, as follows:

"This fundamental fact stands out paramount, that social insurance cannot remove or prevent poverty. It does not get at the causes of social injustice.

"The efforts of trade organizations are directed at fundamental things. They endeavor to secure to all the workers a living wage that will enable them to have sanitary homes, conditions of living that are conducive to good health, adequate clothing, nourishing food and other things that are essential to the maintenance of good health. In attacking the health problem from the preventive and constructive side they are doing infinitely more than any health insurance could do which provides only for relief in case of sickness, and yet the compulsory law would undermine the trade-union activity. There must necessarily be a weakening of independence of spirit and virility when compulsory insurance is provided for so large a number of citizens of the state."[1]

It appears, therefore, that even among those whom sickness insurance is intended to benefit, there is very marked difference of opinion as to its desirability.

SUMMARY AND RECOMMENDATIONS

There can be no question as to the reality or the great magnitude of the burden which sickness and disability now impose on the nation. That one out of every three young men should be unfit for military service and that at least 5 out of every 100 of the entire male population should be infected with a loathsome venereal disease, which is in thousands of cases transmitted to innocent victims, is at once a reflection on national health standards and an indictment of present health conservation programs. An average annual loss by sickness of seven workdays by 40,000,000 wage-earners represents the loss of almost a full year's production for a million workers. The esti-

---

[1]Address at annual meeting of National Civic Federation. New York, November 9, 1917.

mated wage loss of over $500,000,000 per annum from sickness — which takes no account of the incalculable loss due to impaired efficiency on account of illnesses and disabilities that do not result in absence from work — exceeds by a wide margin the total annual dividends of all the railroads of the United States; in a decade this loss amounts to billions of dollars. Yet in the long run the monetary loss is perhaps the least important phase of the problem.

Such conditions call for a vigorous policy of remedial action. Not all of this loss is preventable, but to the extent that it is preventable, the desirability of prevention is undeniable.

That sickness insurance would afford some measure of relief cannot be denied. Sickness insurance, however, proposes to reach only a portion of one class of the population. It makes almost no provision for a great number of disabilities which impair efficiency and it leaves practically untouched the enormously important problem of feeble-mindedness. Yet even in its limited application the annual cost is estimated at the stupendous total of from $700,000,000 to $1,000,000,000; that either estimate would eventually be exceeded is practically certain.

A program calling for any such expenditure would in any case challenge critical examination and compel convincing demonstration of its merit. This evidence is not to be found in the experience of foreign countries where sickness insurance has been tried and where on the one hand it has failed as a preventive agency and on the other hand has placed a premium on inefficiency and fraud. But even if it had worked advantageously in those countries the wisdom of its transfer to the United States where social and political conditions are so radically different would not necessarily follow.

Underlying these considerations is the fundamental fact that all sickness and disability which can reasonably be prevented should be prevented instead of being allowed to remain unremedied until they impose a burden of misery and poverty on the individual and a burden of cost on society.

Preventive work in the case of such communicable diseases as typhoid fever, tuberculosis, pneumonia, and diphtheria has been brilliantly successful. That in less than a generation the mortality rate for tuberculosis per 100,000 population has fallen from over 200 to less than 150 and that for typhoid fever from 35.9 to 12.4 is a tribute to the efficiency of prevention, since in the main these results have been accomplished by preventive agencies. The results already attained with a comparatively modest expenditure in this field are an earnest of the possibilities of still greater progress in the future and of broad success in the field of non-communicable diseases as well, if these are attacked under a definite policy with a permanent and efficient organization and sufficient funds.

The results already achieved in sickness prevention through local effort with limited funds establish beyond a doubt the urgent need for a thoroughgoing investigation of its further possibilities under a definite national policy. Such an investigation should be undertaken at once. The withdrawal from production of hundreds of thousands of the most robust workers for military service has already increased the relative importance of the sickness burden as related to national efficiency, and it will be accentuated by further withdrawals as the war goes on.

Since the occurrence and the results of sickness are nation wide and not local, such an investigation, to be effective, must be conducted by some federal agency, possibly the United States Public Health Service, under the authority of Congress and with an appropriation sufficient to permit of thorough work and conclusive results. The first task of such a national health commission should be to determine broadly the nature, extent, and causes of sickness and physical impairment in the United States, the limits within which a sickness prevention program can be practically applied, a careful estimate of its cost, and a scheme for its actual administration, utilizing, as far as possible, existing agencies already engaged in preventive work but co-ordinating them in such manner as to avoid unnecessary duplication of effort and expense. The commission should submit a comprehensive, analytical, and constructive plan at the earliest date consistent with thoroughness.

It is not the province of this report to outline details of a final plan or even of a preliminary survey. Clearly an investigating commission is bound to take account of such salient features of the problem as:

Reduction in infant mortality.

Supervision of the health of school children, including treatment of various common defects.

Systematic instruction in personal hygiene, diet and living conditions; improvements in sanitation, housing, and milk and water supply.

Extension of free hospital and free clinical agencies.

Treatment of feeble-mindedness and venereal diseases.

The effect of occupation on health of workers.

Periodic physical examination, not only of industrial workers but of the entire population, is another import-. ant matter for consideration. Such a commission may also very properly consider whether and to what extent nationalization of medicine may be advantageous.

So far as strictly occupational diseases are concerned, where the responsibility of industry can be clearly established, these apparently can be most effectively dealt with on a workmen's compensation basis. For this purpose it may be desirable to amend existing workmen's compensation acts to include such diseases or to make provision by separate statute, leaving their administration, however, to workmen's compensation boards in order to avoid new administrative expenditures.

In thus urging a searching study of the merits of a preventive policy there is no desire to prejudice unfairly any proper or necessary measures for alleviating unavoidable sickness and disability or for dealing with individual cases worthy of special consideration. Under some conditions and in a certain field the compensation principle may be justified. But at least it is imperative to narrow that field wherever practicable. The simple fact should never be lost sight of that successful prevention of sickness eliminates the necessity for compensation. Furthermore every dollar successfully and economically spent in preventive work is a dollar invested in productive enterprise.

First a national study of prevailing sickness and its causes; then a national program for the prevention of all preventable sickness, with liberal but intelligent provision for unpreventable sickness through compensation or otherwise, as a duty of society to its members:—this is submitted as a rational, constructive and humane program for dealing with the sickness problem in its individual as well as its social and industrial aspects.

www.ingramcontent.com/pod-product-compliance
Lightning Source LLC
Chambersburg PA
CBHW070802180526
45168CB00004B/1727